Exploring the relevance of Attachment Theory to Dramatherapy practice in the field of Psychosis:

Bringing our Seriously Mentally Ill clients out of the Insane Asylum and back into the real world.

Cane Hill Hospital for the Insane. Surrey. 1974
[where I volunteered as a young teen].

An Independent Study by Rosemary Kate Hughes

Dramatherapy M.A. (University of Derby).

This study is dedicated to my father,

Richard Ronald Humphrey Ellison

(17/12/21 – 12/04/10)

Who, although he did not live to see its completion,

has throughout my life always been my

primary academic support

and advocate.

Acknowledgements:

I acknowledge the support in the production of this piece of work of the following:

My primary supervisor at the University of Derby, Linda Wheildon

My secondary supervisor and support at Manchester University, Katherine Berry.

Karen Courtney, Deborah Birrell and Barbara Jackson at Greater Manchester West NHS Trust who have supported my retraining as a Dramatherapist and all the disruption to normal therapeutic business that it has entailed.

My parents, who have given up both time and space in their home to enable me to work on this project.

John Casson, Sue Jennings, Alida Gersie and Phil Thomas who have provided additional literary/ research material for me to include.

Kim Dent-Brown and the members of the BADth research network, who provided many ideas on the directions my work might take.

David Kennard, Andrew Gumley, Janey Antoniou and other International Society for the Psychological Treatments of Schizophrenia and Other Psychoses (ISPS) luminaries who have shared their work with me.

Simon Bentley, my friend and support, who's very different 'Opus' has been an inspiration in moving forward with my own.

Finally, I acknowledge the contribution of the residents of Tara Buddhist Meditation Centre, Etwall, who have provided many much needed opportunities for quiet reflection during the process of this piece of work.

Introduction and rationale for the study.

My first hypothesis is that Psychosis could be a form of post-traumatic stress reaction; a Psychological defence against the risk of re-traumatisation in which social withdrawal and abnormal perceptions arise from a background of early Psychotraumatic experience and the coping deficiencies due to the resultant disordered attachment this produces. Subsequently when the thus fragile self is later overloaded by inner conflict which threatens to return the individual to an earlier state of fearful impairment, dissociation occurs as a learned protective mechanism. I shall be discussing a variety of new studies that support this view and then going on to look at what this says about therapies that should be offered to this client group.

My second hypothesis is thus that attachment theory has a lot to offer the Dramatherapist working in the field of Psychosis and Attachment-Related Dramatherapy Work, such as that pioneered by Winn, Bannister, Cattanach, Jennings and Lahad, should thus be the treatment of choice for this much-maligned client group. I hope via my own journey through these literary resources to explore perspectives about trauma, attachment, dissociation, thought disorder, imagery, symbolism and Dramatherapy in Psychosis and to test or even prove to some degree my hypotheses around attachment-based Dramatherapy and its effectiveness for this client group. I hope to inform my own future practise and that of others with this piece of work.

This study truly began 30 years ago, when as a newly qualified Occupational Therapist working in the Mental Health Unit of a Children's Hospital, I first met a young man who I shall call Rob. I have changed the names of all clients mentioned in this study and, as well as using only anecdotal details, have altered some of those to add further anonymising in order to protect their confidentiality. At this time a shy and socially anxious pre-teen with an eating disorder who pushed himself into excessive exercise, Rob had lost his father at an early age and his mother, after a period of depression, had remarried an older man of a military persuasion.

We worked primarily using Psychologist, Virginia Axline's Play Therapy (1971) as our primary technique, which Dramatherapists (as shown in Jennings' Remedial Drama' (1978), a book I had been using to inform my work) were already using within their practice with children, and it became clear that Rob's happiest memories were of sporting activities with his father and that he seemed to feel that in order to live up to and retain his father's memory he had to do especially well in this area. However after a few weeks we managed to stabilise his weight and Rob was discharged home.

Then nine years later when I was working at an Adult Psychiatric Day Hospital, Rob became my patient again, but this time with a diagnosis of Paranoid Schizophrenia. He was more withdrawn and in the intervening years his life seemed to have been an ongoing struggle. Rob had not done as well as the family had hoped for in his A-levels and by the time he arrived at my service was very deluded, believing himself to be a superhero of great strength. He ran off his aggression by pounding the pavements around the streets of his home town and heard derogatory voices calling him a wimp and would punch the air and shout out his rejection of these slanders. I met with his mother who informed me that

Rob had consistently found it difficult to meet his stepfather's exacting standards and that she had on several occasions remonstrated with her second husband about the physical punishments he had used to correct Rob's perceived deficiencies.

When I met Rob for the second time it was clear to me that his delusions and hallucinations were very much linked with his negative childhood relationships and traumatic events. By this time I had taken some initial workshops in the use of Dramatherapy in Mental Health and was running a Communication group in which it became apparent that the use of his heroic persona enabled Rob to overcome elements of low self-esteem, as described by Carl Gustav Jung, the founder of Analytical Psychiatry, in 'Man and His Symbols' (1978).

Over the subsequent years of work, I have met a number of clients with whom it was very apparent to me there were similar links and this has led me to pursue an interest in Psychotherapeutic methods and humanistic practice as I have developed my work with this client group and latterly to return to Dramatherapy as a means of combining these philosophies. With this study I would like to look at how childhood trauma affects attachment, how this might be related to dissociation and delusional ideation and to discuss means of addressing these issues through Dramatherapy.

This is part of a new chapter in working with Psychosis – one I am very happy to be involved with – a move towards working with people as opposed to dealing with aliens... The treatment of this much misunderstood condition and of its sufferers has changed much over the past 100 years and especially fast in the past 30 years during which I have been working in the field. Dramatherapy is a relatively new profession which has also developed fast in the past 30 years, but is still only just beginning to be used to facilitate recovery in Severe Mental Illness, rather than being seen as the natural centre of work with trauma-induced Psychosis, which I hope, in some small way, to remedy.

The History of Dramatherapy work in Psychosis.

In 1970 (according to Linford Rees) there was little rehabilitation happening with clients who had the diagnosis of Schizophrenia, other than work therapy/ diversionary activities 'to prevent institutionalisation and promote, where possible, normalisation' (Occupational Therapy), discussion groups, remedial exercise and supported community resettlement - again where possible. Stafford-Clark and Black (1984) state that 'at the end of the 1960s the treatment of most disorders was ill-understood and there had been almost no research on the therapies being used'. This was pretty much my experience when I began work in the unit where I met Rob for the second time.

Roger Grainger, describes the Mental Hospital he worked in at that time as 'an unreasonable, terrifying place,' in which the therapeutic aims of he and his Psychologist co-therapist in setting up a group were: 'to help disturbed minds to achieve a period of freedom from psychological confusion and having to put on a brave face, whilst exploring through imaginative structures the longing for things which could not be contemplated more directly because the emotional material might tear the person apart'.

The final submission of Grainger and Duggan's research project stated that; 'Ordinary life may confirm or disprove the network of ideas we have about the world and if the whole concept has to be readjusted too often, we become finally unable to draw real conclusions anymore'. 'Dramatherapy' they continue 'has a tightening effect on this loose construing, involving as it does participation in an interpersonal experience in which opposing realities are held in relation to eachother and its method validates group perceptions as a trustworthy way of interpreting events, within an environment which is emotionally secure and within which role relationships are clearly defined'.

Ten years later, Dorothy Langley writes that her aims of therapy (at that time) with this client group were: provision of reassurance, stability and a structured, non-threatening environment in which the client can try to assimilate what is happening to him, enabling means of relating to others and helping clients to find ways of coming to terms with their disability and learning to live with its effects via creation of new roles, discarding of old (now unworkable) ones and facilitating expression of individuality which would hopefully improve self-esteem.

She describes small group repetitive movement sessions, trust activities, mime and low-key ball games. At all times she avoids over-stimulation or asking questions and allows clients to come and go should they become over-stressed. This is pretty much how Rob, the client I spoke of earlier, was treated. He was not considered for appropriate for the Psychodrama Group, supervised by Gordon Wiseman, because it was seen as likely to increase his 'acting out' behaviour and the manifestations of his voices, but was involved in some in expressive movement and creative therapy groups. I am not sure that we really helped him much with his difficulties.

In 1981, David Read Johnson depicted his clients as withdrawing from personal relationships and having taken flight into fantasy in order to minimise debilitating loss of self and reported on the work of the Dramatherapist in this field. He discussed the use of creative dramatics to 'integrate the clients sense of identity, diminish boundary confusion and provide structure to manage fear of engulfment by others' and describes Dramatherapy work as having a nature of tolerance of the imaginative which entices the inner self of the Schizophrenic.

He recommended use of improvised roleplay and drama exercises in developing awareness of self and others and of connections between actions and feelings, but discussed the possibilities of further loss of identity and disorientating mental confusion should activities be too complex and suggests ways balance this possibility with enabling socialisation and sharing, concluding with a long process was needed in order to achieve much in the way of development towards therapeutic aims.

Music Therapist, Trudi Schoop (1974) wrote in 'Won't you join the dance?' of her work used mirroring, her body-ego technique, movement and dance to help the patients of Psychiatrist, Eugen Bleuler (who created the term Schizophrenia) to break out of isolation and re-engage with human contact. Schoop's primary aims were: to foster bodily well-being, self-awareness and positive self-view. She pioneered the use of mirroring and imitation, working initially non-verbally, using movement as relaxation and rhythmical breathing as a group. Once her patients had been encouraged to use movement in a fun way, she could help them to build up confidence to use dance and mime individually to promote self- expression and body image techniques to aid communication. Bleuler, she reports, also encouraged his Schizophrenic patients to take part in theatrical events and movement sessions.

Jung (ibid) trained under Bleuler and they shared ideas about their disturbed patients. Jung evolved his theory of the swamping of ego in Schizophrenia and identified the compensatory function of grandiose delusions in relation to deeply-injured self-image, which I saw in Rob. He wrote that auditory hallucinations were an exaggerated form of internal dialogue between different entities within the psyche and advocated use of creative art and writing to explore these with patients, stating that one should nurture the art of conversing with oneself in order to know yourself more fully. He later pioneered his Active Imagination method of encountering the unconscious using visualisation, conversations with inner figures, play with symbolic objects and painting.

Bentall (2004) states that Bleuler believed that repression of painful events could be a part of serious mental illness and identified 4 core symptoms of the newly re-named condition: ambivalence, inappropriate affect [over/under-reactivity], [autistic] detachment and associative loosening, all of which he recognised as stress-related. He viewed the more outward symptoms as secondary and felt that these social and cognitive elements should be the primary focus of therapy.

Whilst Jacob Moreno, founder of Psychodrama, in Who Shall Survive (1933) identified that the more deficient or impoverished the experience of objective reality, the less the opportunity for spontaneity and creativity, the more the child, he proposed, retreats into subjective reality where he is not helpless and neglected. In his view, Psychotic imagery contained the experiences and feelings from which their recipients are trying to escape (as occurs in dreams) and he considered Psychosis as a creative act but one through which social reality is lost and the ability to function in the real world is impaired. He encouraged his clients to embody their voices in order to reduce their power, allowing the ego to gain greater control.

However since then, returning to Bentall: 'features regarded by Bleuler and his contemporaries as fundamental and characteristic have been significantly rejected while those he regarded as accessory symptoms of have risen into ascendancy' and become 'first rank.' Also ideas about using psychotherapeutic principles in the treatment of Schizophrenia, which Bleuler was in favour of, have been allowed to fade... However Johnson's ideas about integrating sense of identity via addressing the inner world of the imagination and the Grainger's regarding clarification and validation of perceptions through interpersonal group experiences continue, in my view, to be valid today. Roger Grainger's more recent work will be discussed later, as will David Read Johnson's subsequent work on trauma.

Exploration of Therapeutic work in Psychosis in the Present Day:

Bentall (ibid) goes on to depict how large pharmaceutical companies have investment in the continued use of their products which may, in his opinion, repress rather than promote engagement in recovery, to demolish ideas about Schizophreniform families, except to say that we have long known that disordered attachment begets further disordered attachment which makes irrelevant, in Bentall's view, the heredity or environment debate. He describes the unreliability of diagnostic criteria as 'no better than astrology,' before presenting grounds for Psychosocial problems being responsible for the development of Psychotic symptoms.

Under the heading Mental Life and the Social Nature of Psychosis, Bentall proposes that the mind-brain adapts to its environment and is literally shaped by experience, as childhood events such as deprivation, abuse or other trauma can result in increased development in some parts of the brain and decreased development in others, resulting in a vulnerability towards developing Psychosis at later date.

Bentall's own work on the Psychology of specific Psychotic symptoms and Psychological mechanisms responsible for them are explored, including studies of auditory hallucinations, exploration of learned cognitive failures which lead the hallucinating individual to misattribute their inner speech to an external source and studies of persecutory delusions, investigating social reasoning, mentalisation and theory of mind biases which lead the deluded person to attribute malevolent intentions to others. His colleague, Gillian Haddock's work on methods of helping hallucinating clients to identify the source of their inner monologues is also discussed and that of Thomas and Leuder, Romme and Escher on learning to live with voices. Bentall concludes that there is no clear boundary between mental health and illness.

Phillips (2005) in 'Going Sane' trawls through literary and philosophical representations of the 'Mad' and the 'Sane' and draws conclusions about changing perceptions. The Mad are variously described as: unconventional, disreputable, catastrophic, implacable, transgressive, estranged from societal norms, saboteurs who endanger society, fallen, judged, improper, undisciplined, broken in their consciousness (Samuel Johnson), 'not us' (which may be the most important, highlighting as it does, the general fear of contamination which permeates society and is perpetuated by today's media), idiosyncratic, difficult, primitive, intense, turbulent, disordered by excess (Charles Lamb), unreliable...

However they are also viewed as: theatrical, passionate, confronting, innovative, enlightened, transformation illuminating, better equipped to tell us about the human condition, poetic, evocative, individualistic, truth-telling, part of all human nature (Melanie Klein), elemental, genius material and multi-talented. They withdraw into the wilderness and return with insights to share/ are transported by their experiences, whose symptoms are a tool-kit for dealing with difficult reality and who challenge complacency. There but for the grace of 'good enough' parenting, he surmises, go us all and thus we can learn much about ourselves through proximity with such people.

Whereas the 'Sane' are represented as: sanitary (this is where the word is rooted), healthy in habits, reasonable, compliant, compelled to conform to societal norms/ spirit-oppressed and consenting to it (R.D. Laing), submissive, humbled, orthodox, harmonious, self-possessed, controlled, reliable, composed at all times, orderly, balanced, law-abiding, complacent, in mastery of their childish and emotional appetites and verbalisations (Erasmus), rooted in stability, acceptable, 'nice,' dull, traditional. They are also apprehensive of their own nature and humanities tendencies towards chaos, plagued by fears of madness or contagion (Charles Dickens), lacking in individuality, self-deceiving, uninspired, uncreative, impoverished in spirit and humanity, having no memorable lines or drama, denying of the existence of chaos (George Orwell) and resistant to change.

Phillips' exploration of the Sane: Mad spectrum concludes that madness and sanity are both part of a continuum and perceptions are interdependent. Disturbance of feelings and constructs, he reports, is part of growth necessary to produce a well-rounded, rather than a one-dimensional, human being and suppression of the realisation of one's own innate self-destructiveness and chaos-production dooms one to mental breakdown in order to achieve the necessary learning to become enlightened about reality.

Thomas and Leuder (2000) also explored the boundary between sanity and madness and concluded that auditory hallucinations; a primary feature for a diagnosis of schizophrenia could from their perspective no longer be considered as such. Thomas (2008) states that we should not assume that reactive thought disorder precludes decision-making ability on day to day issues and is often not long-lasting. While Grainger, concludes that social groupwork helps to reduce it. Meanwhile Larkin and Bloom in 'Dissociation and the Fragmentary Nature of Traumatic Memory' (1996) explored the formation of paranoid ideation arising from a framework of maltreatment and adverse circumstances, calling into question whether the same could be said for delusional beliefs.

Roberts and Wolfson (2004) look at ideas promoted by the Recovery Movement and the Association for Therapeutic Communities in which Psychotic clients are given a significant involvement in decision-making and the practicalities of running their units, based on ideas of collective responsibility, citizenship and empowerment. They argue that recovery-orientated practices increased positive outcomes in Psychosis, quoting 'The Vermont Longitudinal Study of Persons With Severe Mental Illness' (1987) which revealed that over half of clients followed up 30 years later across the state of Vermont had significantly improved or recovered when therapeutic community principles were used and The International Study of Schizophrenia (2001) which showed that over half of 1633 clients from 14 different cultural backgrounds recovered a meaningful life within the limitations of the disorder.

All of the above demonstrate that there is an evolving change of thinking about Psychosis amongst Psychologists, Psychiatrists and the Multi-Disciplinary team, advocating the use of Psychological and Arts Therapies to aid the reduction of thought disorder, to increase socialisation and thus promote positive identity and to reduce the power and frequency of voice-hearing experiences in clients with Schizophrenia/ Psychosis.

To confirm this Current NICE guidelines (2009) state:

Consider offering arts therapies to all people with schizophrenia, particularly for the alleviation of negative symptoms and to assist in promoting recovery. This can be started either during the acute phase or later, including in inpatient settings. or later, including in inpatient settings.

Consider using psychodynamic principles to help patients understand the experiences of people with schizophrenia and their interpersonal relationships.

Aims of arts therapies should include enabling people with schizophrenia to experience themselves differently and to develop new ways of relating to others, helping people to express themselves and to organise their experience into a satisfying aesthetic form and helping people to accept and understand feelings that may have emerged during the creative process (including, in some cases, how they came to have these feelings) at a pace suited to the person.

Considering Post-Traumatic Stress Disorder as a means of looking at Psychosis

The current DSM criteria for Post-Traumatic Stress Disorder (PTSD) includes flashbacks (psychic phenomena), repetitive and/or avoidance behaviour, emotional numbing, dissociative states, generalised anxiety, social withdrawal, feelings of detachment, a pervasive negative emotional state, diminished interest and participation in activities, sleep disturbance, guilt, concentration difficulties, hyper-vigilance and memory problems. All these have been identified at different times as symptoms of Psychosis. DSM Criteria for a diagnosis of Schizophrenia remains the presence of two or more of the following symptoms: Delusions/odd beliefs, Hallucinations/unusual perceptual experiences, Disorganized speech, Catatonia/ Dissociative absences, Negative symptoms (such as restricted affect or avolitional asociality) and social/occupational dysfunction.

Research Psychologists Read, Goodman, Morrison and Aderhold in 'Models of Madness' (2004) contend that while child abuse is acknowledged to have a causal role in adult depression, anxiety disorders, PTSD, eating disorders, substance abuse and personality disorders, surprisingly it is less related to Psychosis and Schizophrenia. Yet the more severe the abuse, they postulate, surely the greater the likelihood of a greater severity of symptoms. They compile data from 65 separate studies over the past 25 years correlating 6 different types of child abuse history with Severe Mental Illness (over half of which identified Schizophrenia/Psychosis diagnosis in the majority of clients studied) and explore the relationship between child abuse and the symptoms of Schizophrenia. Their conclusions are primarily that severity of child abuse is increasingly predictive of adult development of hallucinations and paranoid delusions.

Also Bessel Van-der-Kolk (1995) describes how when people receive sensory input, they generally automatically synthesize this incoming information into narrative form, without conscious awareness of the processes that translate sensory impressions into a personal story. However, he continues, traumatic experiences initially are imprinted as sensory or emotional states, with little verbal representation, that are not immediately transcribed into personal narratives, in contrast with the way people seem to process ordinary information. This failure of information processing on a symbolic level is, he proposes, at the very core of the dissociative experience. These memories are highly state-dependent and cannot be evoked at will.

Van-der-Kolk argues that many of the more complex reactions to early trauma do not readily fall within a straightforward PTSD frame and highlights dissociative phenomena undermining the individual's grounding in the outer (social) world from an early age, hampering reality-testing which makes such individuals vulnerable to subsequent Psychosis because it robs them of internal anchors such as a consistent sense of identity which results in a kind of integral disorientation.

Morrison et al (2003) review research, theoretical writings and investigations into the relationships between trauma and psychosis over the past 35 years. Looking at whether Psychosis can cause PTSD or vis versa and find evidence for both; for Psychosis causing PTSD, they discuss how the terrifying dual occurrence of acute Psychosis and involuntary hospitalisation can produce dissociative effects and symptoms conforming to current perceptions of PTSD and that characteristics of traumatic psychological content lie also within the range of Psychotic experience. It is also known, they state, that PTSD is more likely to occur if the recipient perceives the event as life threatening and delusional ideation can produce this perception.

They quote various studies of Psychotic in-patients and out-patients showing a high proportion qualified for a PTSD diagnosis and a high incidence of childhood abuse and trauma among the Psychotic population, corroborate the precipitating influence of negative life events and/or aversive environmental conditions on Psychotic symptoms, reveal that either a traumatic event or reactivating memories of previous trauma or abuse can engender Psychosis and show evidence that traumatic events correlate with severity of Psychotic symptoms.

While Read, Perry, Moskowitz and Connolly (2001) in their study on 'The contribution of early traumatic events to Schizophrenia' found that patients who reported significant child sexual and/or physical abuse were much more likely to endorse characteristic symptoms of Schizophrenia and also revealed links between the content of abusive experience and that of Psychotic symptomatology, especially commenting or command hallucinations, intrusive thoughts and images of abuse. They identified similarities in developmental neural structure and chemical abnormalities identified as occurring as a result of early traumatic events as also present in adults with a Schizophrenic diagnosis.

Looking at whether the two conditions are both on the spectrum of response to trauma, they suggest that intrusive thoughts and flashback experiences, as identified in PTSD show major similarities with delusional ideation and hallucinatory experiences in Psychosis and identify that hyperarousal/ hypervigilence, disturbed sleep, emotional numbing, social estrangement, derealisation/ depersonalisation, dissociation, difficulties with concentration and neglect are equally present in both, although in Psychosis these are identified as negative symptoms. Again studies also show dissociative symptoms are particularly associated with hallucinations and delusional ideas and that traits thought to be defining of both Schizophrenia and Dissociative Disorder were also present in non-pathological forms in the general population.

Morrison, Frame and Larkin's own study in 2002 of PTSD symptoms, dissociative phenomena and persecutory ideas in Emergency Ambulance workers and Crisis Intervention team members showed that 56% would obtain a positive diagnosis and that preoccupation, distress and conviction of delusional beliefs correlated with a self-blaming attributional style and was predictive of dissociative experiences, whereas a positive view of one's skills and abilities mediated against development of psychotic symptomatology following a series of traumatic experiences. They also discuss burnout as a form of dissociative defence and the cultural acceptability of unusual experiences following recognised trauma, such as those that are part of the bereavement grief process, helping to reduce the distress of the experience, unlike the commonly catastrophic reaction that a diagnosis of schizophrenia produces.

The paper's final conclusion is that all three interpretations of the relationship between trauma and Psychosis have viability and the authors recommend that assessment for both conditions should consider the likelihood of co-morbidity and that the design of services for the treatment of Psychotic conditions should focus on minimising further traumatisation of our clients by reducing the effects by reattribution of disturbing symptoms to an internal/ historical source, using imaginal exposure to and reappraisal of life events via Arts

Therapies and normalisation work such as is recommended by the Hearing Voices network.

Finally Morgan and Fisher (2007) reviewed 22 studies of Psychiatric in-patients with Psychotic disorder and analysed how many of the 897 recipients worldwide identified child sexual abuse and/or physical abuse, finding an overall average on 74%. They reviewed 6 population-based studies in 5 different countries which screened for abuse and/or Psychosis and analysed the results of these, discovering that of over 3 million interviewed, 44% identified abuse history and 18% went on to develop Psychosis.

All of the above shows that in recent years there has been a large body of work on researching the Psychotic experience, describes the current state of research and understanding regarding the genesis of Psychosis and demonstrates the weight of evidence accumulating which indicates that trauma is a large precipitatory factor. All of which would seem to support my first hypothesis that indeed Schizophrenia and its symptoms could be viewed as a severe form of Post-Traumatic Stress Disorder.

Review of the theoretical basis and recognised means of treatment for P.T.S.D.

Linda Winn, works in post hostage situations, with police and healthcare workers, either with individual adults or in groups, using such techniques as narrative, metaphor, story, guided imagery, music, movement, ritual work, sculpting, projective art, enactment and empty chair work to explore emotions and work on communication skills, conflict resolution and stress management and has had success in reducing symptoms, enabling a return to work/life in most cases. In her book (1994) she refers to clients suffering from PTSD as 'liminal people: neither here nor there... betwixt and between leaving their normal lives and being able to return' and the urge to seek therapy as part of having already entered the transitional stage.

Often, she asserts, they describe themselves as being in a jungle, a swamp, a dark cave or an abyss. The therapist who understands her own darkness, fear, chaos and wounds, she offers, can best assist them through their journey, but it is the client him or herself who needs to process the work and review any change in perspectives, not the therapist. This philosophy seems to indicate a background, like my own, of humanistic principles to Winn's practice and also references Jung's wounded healer archetype, which forms the basis of the work of the Alcoholics Anonymous and Hearing Voices Network.

However reactive PTSD, as depicted by Winn, has different features to long-term re-occurring trauma, as explained by Van-der-Kolk (ibid) who suggests that 'Children who have been so traumatised learn to see the world as a frightening and dangerous place and this sense of fear and helplessness can, if untreated, carry over into adulthood: Trauma having effected their primary attachment and thus their ability to develop positive relationships thereafter, those who have experienced Developmental Trauma have not the resources to withstand stress as adults, he maintains.

'These children' he continues, 'have come to organise their neurobiology and Psychology in response to seeing the world as a threatening and overwhelming place and as a coping mechanism to deal with their internal deregulation have used dissociation and fantasy, which as adults they go on doing.' He goes on to describe how trauma in childhood has a powerful relationship to adult health problems which arise, he proposes, due to deficits in emotional self-regulation. 'This results in problems with self-definition as reflected by a lack of a continuous sense of self, poorly modulated affect and impulse control, including aggression against self and others, and uncertainty about the reliability and predictability of others, which is expressed as distrust, suspiciousness, and problems with relationships.

He also identifies alterations in states of consciousness, such as dissociation, depersonalization and de-realisation, flashbacks, sleep disturbances, difficulties in attention regulation, with orientation in time and space and 'being literally are "out of touch" with their feelings. 'These individuals often have no language with which to express internal states and are prone to stress breakdown.' Many of the above symptoms have again been attributed to Schizophrenia, as Van-der-Kolk acknowledges. He recommends theatre work, exploring situations and alternatives to fight/flight/freeze reactions, yoga and breathing exercises, movement, balance work and sensory integration activities, aimed at enhancing self regulation of emotions, reducing hyperarousal and helping them to develop a sense of relaxation and bodily control.

Perry (2001) discusses how traumatised children's brain development is adversely effected by fear, explaining that coping mechanisms usually occur on the hyper-arousal/ dissociative continuum and brain-stem function, which is responsible for more basic, instinctive behaviour becomes dominant over cortex function which controls and mediates instinct. Thus, he suggests, children brought up in an environment of fear may have over-developed non-verbal skills, leading them to misinterpret social signals, and under-developed verbal processing skills, inhibiting learning and development of future coping abilities.

While the majority of homes, communities and schools are safe, he expounds, far too many children experience violence in one or more of these settings. These children must learn and grow despite a pervasive sense of threat. They must adapt to this atmosphere of fear. Young children, he emphasises, are not capable of effectively fighting or fleeing. In the absence of an appropriate caregiver response, the child will soon adopt a model of learned helplessness. In the face of persisting threat, the young child will activate other neuro-physiological and functional responses, such as dissociative mechanisms, involving disengaging from the external world and attending to stimuli in the internal world. This can involve distraction, avoidance, numbing, daydreaming, fugue, fantasy and, in the extreme, narcolepsy or catatonia. He recommends intensive family work and the provision in schools of enriching cognitive, emotional, social and physical experiences which could transform our damaged children.

To summarise this section, the acknowledged experts in working with young people who have experienced childhood trauma recognise that dissociative mechanisms can result in symptoms which in the past have been assigned to Schizophrenia. They emphasise the value and importance of Dramatherapy work such as narrative/ story, bodywork/ movement, projective activities and the use of theatre/ dramatic means to address and work through issues arising from their experiences.

Examining the work of well-known mental health practitioners in the field of trauma.

Babette Rothschild (ibid), whose book's title references the work of Moreno, describes her Somatic Trauma Therapy method: an integrated treatment model of Psychotherapy and bodywork, bringing together mind, body and emotions which helps her clients to 'gain control over symptoms, re-establish a sense of safety, restore emotional resources and nervous system balance.' Rothschild (2003) gives an account of using body memory, which may be retained while conscious, verbal memory of trauma may be repressed, except for unbidden images/ flashbacks/ hallucinations, to explore feelings around safety and security and uses biodynamic techniques to relieve stress. She references the work of Veronica Sherborne in helping a repeatedly traumatised client to use balance exercises to increase body mastery, emotional equilibrium and sense of self, enabling increased communication. She works using an attachment-based therapeutic relationship and regards defences as resources, which can be worked with using spiritual and myth work, gestalt and Psychodynamic methods.

Boris Cyrulnik (2005) discusses how the current social environment which promotes singular striving for personal achievement and self improvement causes, in his view, dissolution of societal bonds which leads to the individual being left in a position of increased vulnerability to the effects of emotional trauma. Resilience, he explains, is developed via internal emotional resources (acquired before or despite the trauma) which are imprinted in biological memory, external resources built-up by caregivers during and post-trauma and feelings regarding the meaning attributed to the event. An individual's experience of reality is dictated by the meaning we attribute to it, as interpreted via dreams and fantasies. Insecure attachment transmitted to the child by a frightened/ frightening parent will impede resilience. However subsequently external models can create meaning in narratives they generate around what happened. The manner in which an event is wrapped can result in pride or shame which will dictate future functioning abilities. This is the work he has specialised in.

Hormonal changes occurring during adolescence can open up possibilities for emotional change, he maintains, and negative representations of the self and others developed in childhood can be modified, but the challenges of adolescence can also precipitate extreme stress and collapse into negative behaviour patterns shaped by unhelpful personal beliefs. Young people who have been bullied or victimised tend to form what he terms 'constructive defences' such as retreating into day dreaming or fantasy imaginings. Cyrulnik works with disturbed adolescents and young adults. A coherent narrative identity that can be reviewed and recognised communally, he maintains, creates a sense of strength and belonging.

Peter Fonagy (1997) links abuse and distorted attachment through the generations in which individuals defensively disrupt their capacity to express feelings and to use self-reflection allowing interpretation of human behaviour, which he maintains predisposes them to experience paranoid anxiety. The reflexive capacity is essential, he states, in order to enable one to make sense of one's own actions and behaviours, and of those of other people with whom one interacts. For individuals in whom this capacity is impaired or reduced, the behaviours of others and their own actions must often be experienced as inexplicably arbitrary and therefore confusing.

When the capacity to mentalise has been prevented or damaged in earlier development, it is necessary for therapists to engage with the patient in a process of enquiry that may lead to mutual enlightenment and, for the patient, the capacity to change. The task in therapy, he proposes, is to hold an emotional conversation. Fonagy also identifies Bleuler's four A's (5 if you count attachment disorder) of ambivalence, inappropriate affect, autistic detachment and associative loosening as all part of Dissociative Disorder.

Sinason (2002) argues that Dissociative Identity Disorder occurs when, in a situation of extreme trauma and helplessness to prevent it, the child's personality splits into different alter-egos or states. Commonly these include the victim, perpetrator and rescuer (often depicted as the three roles of the Drama Triangle), but there can be many more. She specialises in working with people suffering from the range of dissociative disorders and in particular with ritual abuse survivors. She works using Psychotherapy primarily. While Camila Batmanghelidjh, founder and director of Kids Company (2007) describes the case histories of a series of anonymised children she has known through her work who have experienced abuse and gives therapeutic explanations for the effects of trauma experience on abused young people's psychological make-up and behaviours. She uses creative and arts psychotherapies in her work with traumatised teens.

So to summarise this section, Psychotherapeutic practitioners in the field acknowledge that abuse and insecure attachment impede resilience to stress, effect emotional regulation and impair the ability to use reflection, which predisposes them to experience paranoia and advocate the use of therapies which come under the umbrella of dramatherapy work.

Investigating the current practice of Dramatherapists working with trauma.

Ann Bannister describes the evolution of her Regenerative Model, during her 20 years of work with the NSPCC reports on a research project Bannister carried out with children and carers, looking at the interruption of the attachment process in children who have been abused and advocates the use of creative therapies, which follow the developmental process. She discusses Dramatherapy, bodywork, non-verbal, Psychodramatic and play therapy methods used in each phase of her research and reasons for their use, while recognising that the relationship is the key to healing. In the article Bannister she asserts that liminal play space which is essential for creative development, exploration of ideas, testing out of theories and speculation on cause and effect, all of which will need revisiting with children whose attachment has been disturbed and their perceptions of self and other distorted by experiences of abuse.

She discusses Jennings's work on Embodiment-Projection-Role (2003) alongside her of own observations regarding fragmented personality, disordered attachment, difficulties with emotional regulation and distorted self-view in the children she has worked with and emphasises that interaction with the child and engagement with their story in a way that uses metaphor, myth and symbolism to contain frightening experiences and foster emotional control, doubling so the child can watch rather than re-experience traumatic tales and then create positive endings. Bannister also advocates use of mirroring role-reversal and storytelling (which are normally part of a positive attachment experience) to build safety. The conclusions of her research were that self esteem had improved after 20 sessions in the majority of cases.

Gersie (1997) in the chapters in her book on 'Creating a Secure Base' and 'The Search for Re-Narration' looks at heroic stories throughout the ages, stating that there are two types of protagonists: the seeker hero and the victim hero. The former chooses to follow the call to adventure whereas the latter are 'thrown out of their dwellings, by tragic circumstances, into a wide unfriendly world, stunned and ill-prepared for the journey'. This seems to me very much like the experience of a person thrown into Psychosis. Abused children, Gersie asserts, are often 'shocked and surprised to be removed from a cold, unfriendly place, endlessly pondering what has gone wrong and how they are no longer acceptable even to bad people in a bad place' and thus are unprepared to recognise that a different phase of life has begun or to develop any sense of personal agency in the future course of their lives. Often they carry with them this disability for many years, she reports.

Working with these young people, she describes how stories protect from pain, bridge memory gaps, help resolve dilemmas, enable exploration of difficult experiences and personal needs, to be able to overcome defeat and rise again, to develop personal strength whilst gaining support from the group without exposure. She champions the therapeutic group as becoming a 'good enough' therapeutic base from which hope can emerge and recounts looking at home, homecoming and leaving home in storymaking, exploring safety and security, expounding on the need for a narrative which facilitates consent to move forward in life.

When we work with drama in therapy, Cattanach (2008) maintains, we are making a dramatic reality; creating a space set apart, building out of both past and future, a place to begin to examine our fictional world free from the constraints of real circumstance and anxiety, we can escape from being the victims of our real circumstances and commence the journey. In her Dramatherapy group they explore life stages using drama forms and enactments. She also uses projective art, claywork and the Lowenfeld World Technique.

Describing her Developmental Model of Dramatherapy; She states: 'the personal struggle to be authentic means we are constantly exploring the fractured aspects of the self, the buried self, the dislocated self, the self that is not in conjunction with the self we wish to be.' Through drama processes, she reports, 'we can find images, symbols, stories to discover who we are or who we might have been.' She depicts her treatment as starting the journey forward from of disorder, working as companion and guide, with processes enabling clients to advance.

Mooli Lahad, in his article 'A Safe Place' (2009) champions the application of creative approaches in the prevention and treatment of Psychotrauma, describing his Nature Therapy conceptual framework which coupled with creative therapeutic methods can help children develop resilience and support their coping with uncertainty and stress. This method, based on his BASIC PH integrative model which centres on people's natural coping mechanisms in the areas of Belief, Affect, Social Functioning, Imagination, Cognition and Physiology, aims to help people suffering from anxiety disorders and/or traumatised individuals to reduce their symptoms to a manageable level to enable them to regain a sense of control of their lives.

Lahad advocates a space where the child can develop in accordance with his age, with spontaneous playing and creative work at its centre, a place which allows children to express emotions, use the body, to imagine, to be alone and also with the group and emphasises the healing forum of the natural world. These elements, he reports, contribute to the process of developing resiliency.

Also Lahad in Gersie (1996) describes a brief Dramatherapy intervention with an evacuees hostel group in Tel Aviv in which the group were representing their experience of chaos through a communal arts piece and one woman, whose home had been destroyed by missiles and had experiencing choking with the dust as she lay terrified, resulting in a fear of gas masks, became rigid with distress during the activity, which resulted in her breathing becoming impaired. Lahad took her aside to explore her feelings and discovered that all she could produce were criss-crosses of black lines. Whereupon Lahad asked the group to represent these for her and using the group breathing in and out alongside her, she was able to enter the darkness, to become aware of the shades of grey, then with group support was able to unfold the web and follow the lines to lead the way out of the confusion, resulting in much improved breathing.

Frankie Armstrong in Pearson (1996) gives an account of using a lot of breath work in her practice. Deprived and neglected children, whose cries go unheeded, she suggests, lose both their voice (feeling they have no means or right to speak out) and their belief in any ability to influence their environment. It is no surprise, she proposes, that the language is full of expressions such as: 'I just couldn't swallow it,' ' I felt all choked-up' and 'I bit back the words.'

Permission to use the breath and vocal cords to make, alongside others, strong and open sounds from deep within the body is very empowering, she believes, freeing the soul and emphasising the connection between body, voice, psyche and place in society. She uses, with abuse survivors, rhythmical breathing and sound-making, tribal chants, gobbledegook, vibrato, humming, percussive voice, body percussion, improvisational noise and choral song to good effect, then moves on to sharing songs of personal meaning, leading into enactment.

Pearson herself (ibid) worked with disturbed children in a small therapeutic community for children in care with abuse history, using dramatic play and movement with touch. Using touch with abused children can be difficult, she concurs but exercises which enable control are vital to help them to learn body mastery, centring and trust in the self. Pearson recommends using pushing, pulling and leaning, bridge-building and rag-doll roleplay, all of which, she reports, helps children to feel calmer and more focussed. I have been in conversation recently with John Bergman who is currently researching the use of Bobath techniques, balancing work, accelerated breathing exercises and mirroring with young offenders, with the aim of reducing over-arousal in response to stress. This seems to be working with the same aim.

Pearson stresses the importance of boundaries and observes how children who outside the group are wrestling with how to express anger about their experiences will often 'enter the world of story with relief,' choosing a role that allows quietness and softness. This, she infers, may be because the gentler parts of the self have been split off defensively as it feels dangerous to express vulnerability, but in drama the child can let down the barriers between self and others and rediscover these aspects of the self without having to own them. She describes sessions with 'Alison' who was able to find her gentle self within a grandmotherly role, arising from a game of grandmothers footsteps.

Drama, Pearson asserts, is also a way to express anger and hostility through battles with man or beast and strong movement safety, knowing that no-one will get hurt in the real world. Those who feel bullied or downtrodden can choose to be heroes or giants within story and so achieve recognition for their strength and courage, enemy forces can be opposed and overcome and young people can triumph over forces more powerful than themselves, such as Philip who spent some months in individual sessions exploring the story of a Sumo wrestler who learns his skills under the tutelage of a trio of strong women before going on to defeat all the champions in the land.

This reminds me of a young man I worked with in my first dramatherapy placement who we shall call Pat. He had recently lost his mother, been taken into foster care and was having difficulty managing his anger at school. He was also exhibiting episodes of tearfulness, an anxious attitude, was friendless in the playground, easily distracted in the classroom and nervous and clingy with adults. In our sessions strength and masculinity appeared to be important from the start. He viewed himself as too small and clearly aspired to rapidly greater physical stature and success, measuring himself regularly, wanting often to compare himself with various sportsmen and entertainers, to demonstrate his strength and generally to project himself as more substantial and filling of the therapeutic space than he was. He also described his feelings of grief over the loss of his mother as weakness.

Pat seemed to need to explore strength overcoming weakness and did so using heroic swordsplay and karate moves in the roles of the Samurai and of World Wrestling Federation heroes. His feelings of smallness, frailness and weakness were explored via storytelling. Initially he would read and enact stories concerning small characters using a weak-sounding voice and make fun of the characters or act out a BIG monster smashing the character to bits and would laugh, apparently to disassociate himself from any position of weakness. However as time went on he moved on to appreciating stories where the small character used his brains to trick the apparently stronger character into defeat, such as Jack and the beanstalk and the character of Coyote the Trickster.

He used roleplay with these characters on many occasions and through this work was able to learn to express his angry feelings in a more assertive than aggressive manner and to control himself when necessary, such as when our sessions were unexpectedly and precipitously curtailed by school events or circumstances. His foster mother was spoken of in glowing terms until she suffered a personal bereavement of her own and became less available in various ways. For Pat this seemed to emphasise his own losses and feelings of abandonment, but in time he was able to share and re-enact happy memories of times with his mother and acknowledge his distress at being prevented from going to her funeral and seemingly, through identification with his foster mother's experience to begin to say goodbye.

Jennings (2010) emphasises that positive experience creates emotional immunity towards the negative effects of stress. However all this is undermined by experiences of separation, uncertainty, neglect, unrelieved anxiety or pain. Children or adults coming into therapy without these abilities are a great challenge to work with, Jennings acknowledges, but use of Neuro-Dramatic-Play techniques (a refinement of the embodiment stage of her EPR paradigm (2003), involving connections to neural development with the dramatic 'as if' and the built-in human capacity to play), she maintains, can repair damaged attachment when used therapeutically with adolescents and re-enable the capacities for trust, hope, empathy and creative thought.

Attachment deficit, she maintains, in the young person or teenager is revealed through abnormal and often self-destructive ways of trying to be in control. Early patterns of managing distress and expressing anger will have become set for many years and damaged children learn to function in survival mode, resulting in a young person who is isolative, amotivational, mistrusting and lacking in self-awareness, self-regulation and empathic ability, whose limited capacity to express emotion is through rage, depression and boredom.

Often the underlying feeling, Jennings reports, deeply buried, remains one of fear. There are also many suffering young people on the verge of Psychosis who go unnoticed, she adds. However, Jennings believes, that 'as life is a process of continual adaptation, change and development can continue provided the therapist can supply reassurance through constancy and predictability'. She recommends Dramatherapy as the primary modes of intervention for young people with emotional and behavioural difficulties.

However she also reports: 'Although sensory and dramatic play within a context of nurture may well be helpful to many troubled teenagers, for most of them, these activities would seem childish and demeaning. Imaginary work can make them feel ill at ease, mainly because they do not want to look foolish in front of their peers. They cannot afford to 'fail' again and will shore up against intrusion into their fragile inner lives.' Thus masks, sociodrama, quest story enactment, sandtray work, clay, plaster and wood carving, poetry, ritualistic storytelling, rap dance and music are her activities of choice.

Finally David Read Johnson (1995) describes his method of Developmental Transformations, incorporating embodiment, encounter, free improvisation and use of luminal playspace, exploring first spontaneous repetative sounds and movements which, he reports, loosen limiting constructs, then moving into images and improvisation which allows personal images to arise and they can then be explored in a place of safety with the Dramatherapist in role of actor/shaman enacting images for clients as needed and representations of traumatic history can move into those of imaginative retreat and back again, using myth and metaphor, within a witnessing circle which staff and clients can use as required.

To conclude this section, Dramatherapists clearly have a wealth of knowledge and a cornucopia of techniques proven to be helpful in the treatment of those who have undergone traumatic experiences and that the majority of them also identify a link with attachment difficulties, defensive split off of negatively-viewed parts of the persona, lack of emotional immunity towards the negative effects of stress and development of antisocial behaviour patterns which result in isolation, lack of motivation, self-awareness, self-regulation and empathic ability.

The common themes of Dramatherapy work identified are:

Use of movement and sound (Johnson, Armstrong, Pearson, Lahad)

Embodiment activities (Jennings, Johnson, Bannister)

Narrative/ story/ journey work/ metaphor/ myth (Gersie, Cattanach, Pearson, Johnson)

Playspace (Johnson, Bannister, Cattanach, Jennings, Lahad).

Thinking about Attachment Theory and Therapeutic work with disturbed attachment.

John Bowlby, Psychiatrist and Psychoanalyst, and Mary Ainsworth, Research Psychologist, define attachment in 'Child Care and the Growth of Love' (1965) as the basic need and strong disposal of a child to seek contact with and proximity to a specific primary care-giving adult, especially in times of anxiety (even if the attachment figure is the cause of the anxiety) and unaffected by situation or the passage of time. Attachment behaviours are identified as those engaged in to obtain or maintain the desired contact or proximity but are also exploratory behaviours enabling outward-looking, social engagement and the development of self-identity.

Bowlby and Ainsworth describe the urge for attachment as having a psychological purpose, because positive attachment develops the brain and imprints on the young mind a model of positive relationships. From the early months onwards the presence or absence of a strong attachment figure determines, they inform us, whether a person is able to respond appropriately to any potentially alarming situation, can form social relationships which will provide support and it is also within close attachment relationships that children learn to make sense of themselves, other people and social interactions. If they experience that they are loveable and accepted by others, they become confident and self assured. However separation(s) from the primary carer, neglect or lack of affection/nurture may have grave and far-reaching effects on character and so on the whole of a child's future life.

Jeremy Holmes (ibid) in 'The Search for the Secure Base' (2001) argues that attachment is the Psychological immune system, describing self-esteem and security as intimately linked with regulation of affect and a capacity for thinking about relationships, which optimises our resource-holding potential. However, if attachment bonds are weak, he reasons, the system can react excessively to threat, resulting in fear of intimacy and catastrophic thinking, both of which can be devastatingly emotionally disequilibrating. Failure of the normal differentiation between self and other is central to Psychotic illness, he reports. In Paranoia, sufferers may attribute to reality malevolent intentions that that properly belong to the self and in Schizophrenia the normal labelling system by which we distinguish our own thoughts and feelings from external perceptions is compromised. As a result these unfortunates feel defenceless and may react to psychic phenomena with extreme terror, withdrawal or occasionally violent counter-attack.

Looking particularly at the chapter on 'Abuse, Trauma and Memory,' in which Holmes discusses how children in the absence of a secure base, have their capacity to develop a self-reliant and positive sense of self compromised and in the case of abuse, the very coherence of narrative thought patterns may be disrupted and the ability to distinguish reality and fantasy may be confused. Thus, he writes, we encounter clients who are overwhelmed by hyper-realised memories which they cannot quell or conversely people who are suffering from repressive loss of memory regarding much of their childhood. He emphasises creative exploration, particularly art, creative writing, poetry, claywork and use of story in working with such individuals.

Susan Gerhardt, in her book 'Why Love Matters' (2004), describes development of the infant brain as 'experience dependent.' A child's whole being, she argues, tells them to approach, trust and rely on their parents, but when experience tells them that those same people may hurt or frighten them, fall into the 'approach/avoidance dilemma', resulting in insecure attachment. This, she maintains, is due to the trauma having effected their primary attachment and thus ability to develop positive relationships thereafter and also that the trauma persisted during a key time in their development impacting not just on their emotional development but also their cognitive and neurobiological development. Stress, she goes on, occurs when high arousal is perpetuated either by lack of respite or insufficient time recovery between stressful events.

Hyperarousal overwhelms the normal homeostatic mechanisms and the brain puts on hold permanently other cognitive mechanisms, such as the ability to learn new information, to relax, to access memory, to reflect, interpret and thus adapt, to process social cues and the body in constant preparation becomes weary. 'These are also the children' she states 'who are most at risk of developing serious Psychopathology in adulthood. The earlier a child experiences abuse or neglect, the smaller the brain volume, particularly of the pre-frontal cortex which is so vital in controlling and calming the emotions. Some go on to become diagnosed as suffering from Borderline Personality Disorder or Paranoid Schizophrenia.'

To further review Attachment Theory, I return to Sue Jennings (2010) who discusses in her chapter on the subject of NDP and Attachment, The importance of John Bowlby's thinking cannot be underestimated in the field of attachment, she states. Bowlby, she submits, 'helps us understand the importance of mourning in trauma and loss or indeed the effect of the disrupted attachment, in our work as therapists.' Jennings then reviews Mary Ainsworth's work on insecure styles, such as ambivalent, avoidant or disorganised attachment and goes on to identify children she has encountered in her practise who fit the descriptions. Children who have been traumatised learn to see the world as a frightening and dangerous place, she reports, and this sense of fear and helplessness can, if untreated, carry over into adulthood, affecting their ability to develop positive relationships thereafter. 'When secure attachment does not take place', she writes, 'then such children will have difficulties in communicating appropriately, feeling empathy and forming relationships'.

Bloom (1996) proposes that any traumatising event produces a major disruption in the victim's core beliefs regarding personal safety and causes a profound physical response, which consciousness tries to repair by separating emotional connection from the event and disengaging verbal memory. However the trauma continues to wreak havoc in the unconscious sphere and is recalled via visual, auditory, kinetic and bodily means, as the traumatised person experiences unbidden and re-traumatising images, sounds and sensations. In order to treat traumatised people, she thus suggests, one has to access the non-dominant creative hemisphere of the brain and for this theatre work is ideal.

Human beings are natural mimics, she explains, and ritual is an ancient voluntary group mimetic process which consolidates security via unification and demonstration of shared beliefs. Theatre evolved from rite of passage rituals and also shares elements with play, but unlike play may not have unhappy associations. Movement is another natural process and so use of music and dance can help a client to access positive feelings. Thus with her clients she begins with bodywork then moves into the verbal sphere, thus bringing back together feelings and thought and overcoming learned helplessness through healing ritual. Then she uses story work and enactment of myths, using the metaphor of journey.

In conclusion, Attachment Theory provides a theoretical basis for understanding some of the complex difficulties that this client group has in interpersonal relating and provide the defining features necessary for Dramatherapy, such as the requirement for a secure therapeutic base, essential to the working alliance needed to bring about therapeutic healing and on the role of the interaction between therapist and patient. It emphasises the importance of affective processing, especially of grief, loss and separation and recognises how patterns of insecurity are replicated and perpetuated, not only through the life cycle of the individual, but through the family life cycle from generation to generation.

Common themes in addressing attachment disorder via Dramatherapeutic means are creative/ projective work, use of story, movement, bodywork, music, rituals, myths, metaphor and journey. All of these were also shown to be used by expert Dramatherapists in work with trauma in the previous chapter, which I find unsurprising as traumatic experience and attachment disorder have been revealed to be reciprocally interactive.

Linking attachment-based trauma work with Psychosis.

Berry and Barrowclough (2008) found that avoidant attachment correlated positively with paranoia and positive symptoms and also correlation was found between attachment anxiety and hallucinations. Also Stevens and French (2009) found strong relationships between emotional abuse in childhood and hallucinations, delusional ideation and thought disorder, between emotional neglect and delusional ideation, between sexual abuse, hallucinations and thought disorder, and between physical abuse, dissociation and all three symptoms.

Berry and Barrowclough (2009) later reviewed adult attachment, perceived earlier experiences of care and trauma in people with psychosis. They found indeed a negative correlation between reports of parental care and avoidant attachment and a significant positive correlation between reports of overprotection in care-giving relationships and anxious attachment as well as higher levels of attachment anxiety in patients reporting trauma involving significant others in childhood relationships, compared to patients reporting trauma involving significant others in adulthood.

Thomas (2008) proposes that voices are associated with the hearer's personality and the life stresses in his/her past experiences and life history that have produced them and advises that practitioners not assume that Psychosis compromises the ability of our clients to engage in narrative work. He suggests that recovery involves the act of reclaiming language, owning and sharing stories and having an audience for them. Yet the dominant view is that Schizophrenia leads to such a profound loss of selfhood, that recovery through narrative is generally considered implausible.

So how, he asks 'are we thus to interpret the absence of narrative that appears to be a feature of Psychosis?' Thomas suggests that clients with Schizophrenia have been considered without ability to work decisively towards their own recovery as they have been viewed as having an impoverishment of persona. However, in his view, this loss of perspective, during an episode of Schizophrenia, is merely temporary and when the acute phase passes the exercise of trying to bring meaning to the experience can begin.

Experience, he contends, is immediate and thus needs a participatory response before one can reflect and find meaning, particularly if the experience is ongoing and if the person feels under threat. Thus narrative failure may be due to an inability to articulate what has occurred. Often, he maintains, words are inadequate to convey the experience of Psychosis and thus the individual needs to turn to other creative means to be able to return to 'being in the world.' A traumatised client may need to retreat to an internal place of safety and may use ritualistic behaviours to enable gradual reintegration and reconstruction of life roles and rebuilding of self-esteem before they become ready to face interpersonal interaction. Narrative loss, he posits, is only a symptom representing the intensity of fragmentary experience that thwarts integration. Thus we should move away, with these individuals, from the 'hegemony of language' to other forms of self-expression, such as dance, music, performance, art and poetry.

Gumley (2009) proposes that individuals who as a result of early adverse experiences develop a dismissive or avoidant attachment style, evolve into personalities associated with poor pro-social coping and help-seeking strategies, more negative self-view, less heightened sensitivity to interpersonal stress, impoverished meta-cognition and limited mentalisation processes, such as reflective functioning. Thus such personalities are both more likely to develop paranoid ideation and have dissociative experiences when subsequently subjected to stress and are less likely to be able to identify personal goals which are needed for recovery from serious mental illness.

The degree of attachment security attained in childhood has profound influences on a person's ability to conceive of or imagine mental states, both in oneself and others and is related to the concept of 'self-reflective function' and 'empathy', including affective as well as cognitive elements of mental processing, which develop later as the child matures into an adult. Key barriers to recovery from Psychosis, he asserts, include perceptions of shame, social anxiety, self-criticism, hopelessness; all of which are all associated with abuse history. Thus, he emphasises the importance of a therapeutic alliance enabling the reaching of a collaborative understanding of the individual's personal response to emotional distress and working towards development of affective management and of reflective function, using any accessible creative means.

Finally, Read and Gumley (2008) pull together some of the above and other studies in order to discuss whether attachment theory can satisfiably explain the relationship between childhood adversity and Psychosis. They look at the statistics regarding the development of Psychosis and poverty, deprivation, urban living, immigration, sexual and physical abuse, emotional abuse, neglect, parental loss (Bleuler found as long ago as 1978) that 31% of 932 people diagnosed with Schizophrenia had lost a parent before age fifteen), bonding dysfunction and disordered attachment. Social causes of Psychosis and disordered attachment, they report, have been investigated in a range of studies over the past 10 years, which are reviewed.

To summarise the last two sections; It is clear that recovery from Psychosis is possible and that those undergoing it value Psychological/ Psychotherapeutic input, creative writing, storytelling, art therapy, a spiritual focus, relaxation, dance, music, performance and poetry. Links between trauma, disordered attachment and Psychosis have been revealed and health professionals also advocate Psychological/ Psychotherapeutic/ Psychodynamic therapies be used in the treatment of Psychosis. Psychotherapy and the Arts Therapies have long been used to treat adult difficulties with attachment, as have narrative, drama and story, enabling exploration of personal historical events and the sense of self in a safe manner.

Looking at existing Dramatherapy work with clients undergoing Psychosis.

Dent-Brown and Ruddy (2006) analysed all randomised trials that compared Dramatherapy and related approaches with standard care or other psychosocial interventions in the treatment of Schizophrenia, where outcomes had been measured (there were only five that met this criteria, 210 in-patients were studied in total and the trials spanned 28 years) and concluded that although they could cautiously acknowledge small improvements in social function, self esteem and reductions in feelings of inferiority and in negative symptoms, these were not in statistically significant numbers and none of the studies had involved any long-term follow-up, thus they were unable to conclude any substantial benefit in Dramatherapy approaches.

Yotis's paper (2007 reviews the work of David Read Johnson with Psychotic clients at Yale; his investigation of his clients boundaries and relationship between self and other, using improvisational role-playing to explore communication styles (1981) and subsequent examination of the degree of fluidity and rigidity in social role representations during role-play among paranoid schizophrenic, non-paranoid schizophrenic and control groups (1988), concluding that patients evidence strong efforts to structure and overly define boundaries in their role-playing, whereas non-paranoid patients demonstrate more fusion and merging of boundaries, and clinical trials indicated that theatre performance in psychiatric institutions improved social contact (1990).

Other studies, Yotis reports, attempted to measure the impact of Dramatherapy on specific functions of the disordered individual, concluding that drama activities resulted in an increase in social interest, mood elevation and increased the frequency of verbalization of group members, but like Dent-Brown he concludes that these findings were not statistically significant.

Yotis finally analyses his own doctoral research study in 2002 which examined the impact of therapeutic performance in Dramatherapy practice for clients with Schizophrenia, using the Dramatherapy Performance Evaluation to measure improvement in client's relationship to self and others, the outcomes of which, he asserts, showed the importance of drama's unifying cathartic structure, demonstrated how non-verbal processes reinforce the impact of verbal ones and showed that Dramatherapy performance had indeed a significant effect on clients' dramatic involvement within the group process, decreased their overall negative symptomatology, increased self-esteem, competence and perception of support from carers and friends..

Emunah (1994) presents the case study of 'Lisa', who has a diagnosis of paranoid schizophrenia and at 27, she describes as waiflike. Lisa is seen by nursing staff as rigid, controlling, isolative and unemotional and doesn't involve herself with other patients. At times she seems almost catatonic. However in Emunah's Dramatherapy group she becomes unexpectedly imaginative and creative. She enjoys improvisation, but can go from moving gently into violent expressions of apparently unprecipitated fear or anger, abruptly started and stopped, causing for the group, as Emunah describes it 'trauma with no follow-up.'

Lisa's history is of spending her childhood in a series of foster homes, after her mother suffocated her twin brother in room next to her when she was aged five and as the group progresses she begins to suspect that her ability to cut off her feelings commenced with her brother's death. Emunah uses roleplay, exploring group member's day-to-day situations and Lisa is happy to play roles for others, but feels that her own life is insufficiently interesting for dramatic portrayal.

Yet when the group moves on to core life issues, utilising Psychodramatic techniques, Lisa asks if she can revisit her brother's death and with the group's help, plays out a sequence of scenes in which her mother enters her brothers room stating she will read him his bedtime story, Lisa then hears choking, sees from the doorway her mother suffocate her twin with his pillow, then her mother comes out of the bedroom crying and calls the Fire Dept. who come and take both her lifeless brother and now hysterical mother away and Lisa never sees either of them again.

Instead, she tells the group, Social Services sent over her first foster mother with whom Lisa left the family home behind her. No-one asked her how she felt (it would seem that indeed her extreme trauma had got no follow-up) and she changed from an effervescent kid to someone who used books, listening to music and solo-play to avoid relationships. Emunah creates a scene where a caring adult was able to listen to the five year old's feelings and Lisa finds the words 'I want my brother back' and then through role reversal allows Lisa to own her emotional pain, fear and confusion. They explore whether she could have done anything to save her brother and her fear that had she interfered she too would have died. Lisa is able to reassure her five year old self that she was not to blame. She goes on to become more expressive generally and this is also observed by nursing staff outside the group situation.

Augusto Boal, Dramatherapist, in 'The Rainbow of Desire' (1995) depicts how a baby first discovers fear and anger when he is hungry and finds the maternal breast is not available. He goes on to expound how the child's freedom of expression is constrained by adult dictates who enforce what we become and so reduce the potential personality. Boal describes his Cop in the Head technique as dealing with internal dictates that obstruct the will, foster passivity and oppress the natural desire to express oneself and act creatively and thus to grow emotionally, develop our personalities and self-actualise. Through repulsion, hatred, fear or violence ideas penetrate the mind and the psyche by a kind of social osmosis. He recommends physicalisation of internal voices, enabling addressing of the issues that oppress. When the oppressed creates images of his/her own oppressive reality, he plays with the images he creates in the aesthetic world in order to modify the real one.

This technique seems very applicable to those who hear their abusers voice constantly telling them that they are worthless as has been the experience of many of my Psychotic clients. Again I shall change the name and some of the details of one of my more recent clients who heard the voices of his physically abusive uncle, taunting cousin and a voice that he assumed was from God, encouraging him to make himself more useful in the world. Let us call him Sam.

This gentleman experienced a difficult childhood: he had been born with slightly webbed hands which had resulted in his being teased at school throughout his young life and also beaten at home by his alcoholic uncle, who held the family in his control. He came to my unit after a Psychotic breakdown resulting from an incident in his first job where colleagues had encouraged him in making a fool of himself in front of management and subsequently he had been taunted for his foolishness, which had precipitated his voices.

Sam was extremely incapacitated by these voices, the occurrences of which caused him to freeze, break out into a very noticeable cold sweat and to either read and re-read any available writing or repeat song lyrics over and over under his breath in attempts to drown them out. In the Communications Group (a weekly loosely-dramatherapeutic assembly with aims of improving social skills and confidence) he demonstrated a formidable intellect but severe self-esteem problems, which he was unable to overcome in order to pursue social, educational or employment opportunities and also his voices would accelerate should he try to do so, telling him that he was a no-hoper who no-one would wish to befriend or employ.

During Sam's admission I set up a Coping with Symptoms Group with the aims of promoting awareness of voices/visions and unusual thinking in the context of normal experience and emphasising that voice hearers can live meaningful, productive lives. Sam attended the group within which I acknowledged some resonances with our client's experiences, such as having an internal negative 'voice' in the form of learned negative self-statements and periods when stressful life experiences blew out of proportion one's ability to think clearly. Following this group, Sam was able to reduce the regularity and power of his voice hearing experiences and to begin to move forward with his life.

In his book (2004) Casson considers the efficacy of Dramatherapy and Psychodramatic interventions for voice hearers and discusses how dramatic aesthetic distance makes therapeutic work safe for these much traumatised clients. He describes the voices of his clients as falling into distinct categories: Voices of unknown people (which can be positive or negative), voices of known people (tending to be negative, often abusive), commentating voices (usually negative), supportive voices (positive), spiritual voices: Gods, demons, archetypes and power figures and whisperings and noises (usually interpreted as negative).

Voices, like abusers, he asserts, breach personal boundaries, they are invasive, intrusive, distracting (preventing concentration, increasing stress and disrupting communication. 100% of John's clients disclosed trauma in their history: 67% had experience physical and/or sexual abuse and/or rape, but the other 23% had survived racist assault, severe bullying, near-death, terrorism, genocide, the dislocation of refugee status, domestic violence and one had witnessed a murder. Casson gives a range of examples of client's experience of the voices of abusers/ oppressors resulting in feelings of powerlessness and/or worthlessness and sees a clear link between voice-hearing experiences in his clients and low mood/ isolation.

In his book John puts forward his Environmental Model for the genesis of Psychosis where negative forces accumulate and attack the vulnerable self/ego until the individual eventually is unable any longer to process the trauma, becomes overwhelmed and withdraws into alternative reality, which is both a haven of safe retreat and a hell inhabited by fantastical creatures. Theatre which inhabits the place of the psyche, illusion and myth, he contends, is the natural forum for the journey back.

He describes working through the EPR (Jennings: 2003) process with his clients, starting with Tai Chi to foster relaxation and promote a sense of power and bodily control in co-operation with others and voicework, explaining that survivors of abuse can rarely relax lying down and may have breathing difficulties associated with fearful breath-holding. In the Projection phase he used imagined landscapes, spectogram and world techniques, storytelling, journey maps, ritual, myth, work with texts and poetry and in Role puppetry, videowork, babushkas to explore the parts of the psyche, visualisation and enactment, photo therapy, mask and costume, brief psychodrama to re-enact their dreams and nightmares and role reversal to explore voices and their meaning.

Casson provides clear evidence that self-esteem improved in his client group and the intensity and negativity of voices decreased, clients developed new communicative abilities, became more assertive and able to voice feelings and opinions and less isolated and more relaxed inside and outside the group situation, as is endorsed by the staff he worked with. Clients also gained some insight into the sources of voices and found talking about them liberating. Working through issues enabled owning of voices and previously split off feelings and discovery of increased hope for the future.

Dunne (2003) describes her work using Narradrama (a method of working with personal narrative using Projective activities, enactment and sculpting) with disturbed young people: 'Crystal' uses material and puppets to explore her inability to trust others, her disconnectedness, vulnerability (personified in a lamb puppet), confusion, fears of darkness, further abuse and exploitation and her negative voices (embodied by a stuffed rat) which lie to her about her abilities and paralyse her will. She manages to cloak herself in self protection and kick the rat across the room.

Later, again using the image of a lamb, she explores via story fears of attachment, of trusting her feelings and her tendency to lie to herself and others. She shares wanting to help others who have been abused but experiencing sight and hearing fading out when she feels overwhelmed by others pain and vulnerability, touching on her own. By the end of the group she shows via the closing ritual that she feels that she is on the way to controlling her fears and is in the process of mending her hurt self.

'Isabelle' depicts a significant scene in her childhood when her mother tried to kill them both by jumping off a bridge and uses a sparrow to represent her guardian angel who she feels saved her when her mother eventually changed her mind after the child's coat fell into the water. She shares with the group that she has ever since been fearful of bridges and water and discusses how spirituality has been her strength. Later she further explores her relationship with her mother who she depicts as a tall tree who is impossible to reach up to and later still writes and reads a poem about herself and her brother in which the world spins faster and faster and becomes a chaotic blur, but together they manage to slow time down. Towards the end of the group she draws her sparrow and writes about the lies she discovered when little birds told her the truth and the cats that have to be avoided. By the end of the group she uses the closing ritual to portray her determination to achieve a phobia-free future.

Win Browne, Dramatherapist, and Alison Clarke, Artistic Director of 'Dramatic Results' and 'The Stop Gap Therapeutic Theatre Company' have provided me with the write-up of their experimental Forum Drama work over the course of eight weeks with young women prisoners at HMP Send, Woking. They used creative writing, storytelling and drama using fictitious characters, sociodrama, videowork, music, poetry, visual artwork, phototherapy, puppetry and spectogram figures to enable clients to re-enact episodes from their history within a supportive group, explore their different voices and express their needs, goals and concerns.

The group of prisoners all had a Dual Diagnosis of Schizophrenia and Substance abuse and also 30% of their group had history of child abuse, 40% harmed themselves and 60% had suffered domestic abuse. They worked together towards a theatre piece, which would be screened outside the prison, alongside an exhibit called 'Writing on Walls' which was exhibited in a public place. The verbal and written responses of women recipients were collected and then brought back inside, before a women's creative writing group from the community asked to visit the prison to hear the prisoners who had worked on it describe their experience, which resulted in some ongoing correspondence between members of the two groups and improved the self-esteem and motivation of the prisoners

Meldrum (2006) presents the case study of 'Beryl' who, like Emunah's 'Lisa' cannot feel anything, having married her childhood abuser and, she feels, caused her Mother's death in the process.. Using creative writing, story, text, roleplay and through exploring her relationship with Meldrum, Beryl finds solace in attachment theory, is able to own her love for her mother, and leaves both her elderly husband and her guilt behind. In conclusion, Meldrum describes her role as a Dramatherapist as to help the client to reflect and think about themselves, which allows them to regulate their own emotional reactions and also to develop the client's capacity for empathy, by assisting their getting in touch with their own feelings and facilitating reflection on how these affect them and also impact on others.

Grainger in his 1990 research paper on 'The Effect of Dramatherapy on Schizophrenic Thought Disorder' explores whether drama can provide a field that is safe to practice in, where personal relationships can be reviewed and disordered ideation modified, concluding that 'thinking was more ordered and there was evidence of the integrating effect of drama upon fragmented human awareness' after a brief project, even though he used only a small client sample.

Also Grainger (1995) in the chapter 'Drama and Schizophrenia' describes Schizophrenic patients as 'trying to make sense out of an arbitrary and unpredictable universe' as having been subject to inconsistency, the constructs he/she develops to comprehend the world share these characteristics. Faced with this situation over a prolonged length of time, he proposes, 'the individual adapts by withdrawing into a fantastical world in which thought and feeling can be more straight-forward and/or by reproducing ambiguity and irrationality in his own thinking and behaviour.'

Schizophrenia, he maintains, originates from metaphysical insecurity and leads to existential confusion, where the sufferer constantly misconstrues what is happening and the intentions of others, having been systematically prevented from forming a reliable belief system. In such a world, he extrapolates, nothing is expected or unexpected and understanding becomes provisional. When working dramatically with thought-disordered people, he emphasises, the primary therapeutic goal is to reverse the spiral of communicative breakdown, increasing confusion and defensive retreat from reality.

To do this one needs, he submits, to provide evidence that the group environment and the people within conform to a comprehensible model of being and that the structure of common reality can be upheld. Dramatherapy as an interpersonal experience, he proposes, is particularly good at tightening loose relationship constructs and providing cognitive clarity through participation in the essential processes of human experience and organising concepts regarding relationships via role-play, scene enactment, shared exercises and comparison of interpersonal similarities and differences.

Snow (in Gersie 1996) in his chapter 'Focussing on Mythic Imagery in Brief Dramatherapy with Psychotic Individuals' concurs with Campbell's description of ' the spontaneous fantastical symbolism of the Schizophrenic patient who has fallen backwards unprepared into an inward journey haunted by mythic imagery' rather than the prevailing Medical Model, which he states views Psychotic beliefs and experiences as meaningless primitive process malfunction effects, rather than (as he does) the building blocks of a potentially transformative process. He discusses the ideas of Jung, Rebillot and Weir Perry (ibid) on whose renewal through dramatic ritual he bases his practice and introduces his 'Psychic Renewal Process' through which the client progresses via regression, disintegration and resurrection.

Snow shares his own experience with the client group, in the form of a case study on Peter, aged 39, who when Psychotic saw himself as a saviour of a doomed world and had written a play in which Snow recognises Perry's sequence of archetypal motifs and elements of the heros journey. The group work through the story improvising on themes, images, myths and fantasies and Peter is able to share and investigate parts of himself, achieving a closer relationship with reality. However details of the process are not clear.

James Roose-Evans, Theatre Practitioner, writes on personal ritual (Journeys of the Heart) in Pearson (ibid) looking at rites of passage for parting. He states that society has no satisfactory ritual for leaving a broken home or relationship, surviving abuse, becoming a woman or going through the menopause, recognising the death of childhood... coming to terms with reality... reinforcing the ability to cope with an unpredictable world, surviving generally... and discusses his work with young people in a Mental Health Centre.

Ritual, he suggests, enhances our capacity to move forward, develop, live... by re-awakening and validating deep layers of the psyche. He uses journalling alongside rite of passage work to help clients process their journeys which they then share with the group, for example 'Christian' describes how he was able to forgive his mother by understanding that she also had been trapped during his childhood and process his own fears about repeating her patterns, how 'Monica' explores entering womanhood and anxieties about her developing sexuality and 'Dixon' rehearses his marriage, challenges the lies he tells himself about perfection and comes to accept that another person cannot take away all ones fears of loneliness.

In my second Dramatherapy placement I case-studied, Max, a 36 year old man with a diagnosis of Paranoid Schizophrenia, the father of 3 children, separated from his partner and living in supported accommodation. When I saw Max individually prior to the start of the group he portrayed his voices as parts of himself and his social network was pretty sparse: in order of closeness he put his children's mother, with whom (he stated) he had some contact, his father facing out and he placed his children at the edge of the circle, with the eldest looking after the youngest.

In introducing himself to the group via the persona of a friend, Max took the role of his fellow-parent to their children, Adele, and was able to say some positive things from that role about Max as a father, to which the other group members responded supportively. He also spoke about wanting to express to his Psychiatrist his fears of a depressive relapse, but also needing to be seen as capable of achieving positive self-direction. For this dilemma he received support from the group, and was then able to ask for support in dealing with his increasingly negative feelings about himself, stating that he had a strong self-punishing side.

Early on in the group Max's scene of angrily confronting his Psychiatrist about wanting to feel in control of his own life, was acknowledged as particularly powerful by the whole group and Max was able to respond to this by stating that he felt his frustration had been understood and to share feelings of being over-monitored in his sheltered accommodation and anxieties about being viewed negatively by decision-makers who could affect his progress, living situation and access to his sons which he hoped would be facilitated soon.

Midway through our sessions, I spread out a series of pictures of faces showing possible different emotions and suggested they select whichever they felt some connection with, then discussed these in small group. Max chose a picture of a man looking contemplative and it became clear that it had put him in touch with some strong feelings about himself. When I suggested that he use an empty chair to talk to this part of himself, Max informed me and the group that he was spending too much time in this negative company already, but with support was able to connect with a more self-nurturing, spiritual part.

Max later spoke about a part of himself who tells him he'll be laughed at and rejected if he tries to go out and mix with people and encourages feelings of self-blame over uncomfortable social situations. Many of the group felt that they too had an undermining part of themselves which attacked their self-confidence. We used the sculpting and doubling and Max spoke to his negative self, stating 'you need me but I don't need you' and received encouragement from the group.

Towards the end of our sessions the group explored what blocks us from being the person we wish to be, by first using visualisation to create a block, then finding a space and pushing against it with eyes closed, then becoming the block and experiencing it's presence when placed between the present and hoped-for self. Max recognised that his block was internal and took some space to explore his past and future role as father and family man. the evaluation session afterwards, looking back at his aims, he stated that he felt he had moved forward on identifying and expressing his feelings and in self-confidence overall.

In conclusion, I feel that I have shown in this last section, through descriptions of the work of other Dramatherapists in the field and also looking at my own case histories that addressing traumatic experiences and voice-hearing via Dramatherapy processes is a viable way of working with trauma-based Psychosis and these are exciting new ways of working with the client group. Casson, Browne, Emunah and Dunne all recognise that trauma is at the base of their clients' difficulties and are working with it. Use of improvisation, myth/story, sociodrama and Psychodrama have not until very recently been considered appropriate with Psychosis and work with voices/ personal history/ emotional journeys and identity as demonstrated by the work of Casson and Roose-Evans and is pioneering, as is Casson's application of the embodiment-based work of Sue Jennings, David Read Johnson and Ann Bannister to this client group.

Summary of the findings of this study and conclusions.

I have shown, I believe, that thinking about Psychosis is moving towards promotion of the Arts Therapies in this field of Mental Health (NICE Guidelines: 2009) and that there is now recognition of its use in working with thought disorder (Grainger: 1995) and voice-hearing experiences (Casson: 2004), as well as with negative symptoms (NICE Guidelines: 2009). I have described the current state of research and understanding regarding the genesis of Psychosis and have revealed evidence (Haddock: 1993, Larkin and Bloom: 1999, Bentall: 2004, Thomas: 2008) showing that childhood trauma is believed to be a large precipitatory factor for the condition.

Differences and similarities between Post Traumatic Stress Disorder, Dissociative Disorder and Psychosis have been explored (Winn: 1994, Van-der-Kolk: 1995, Perry: 2001, Romme and Escher: 1999) and the three conditions established as being linked (Bloom: 1996, Morrison et al: 2003, Morgan and Fisher: 2007). Links between trauma, disordered attachment, cognitive mechanisms and Psychosis have also been revealed (Romme and Escher: 1999, Berry and Barrowclough: 2008, French and Stevens: 2009, Morrison et al: 2003, Gumley: 2009) and I have demonstrated that the acknowledged experts in trauma work recognise that defensive dissociative mechanisms can result in symptoms which in the past have been assigned to Schizophrenia (Van-der-Kolk: 1997, Fonagy: 1997, Holmes: 2001). Also trauma expert therapists have been shown to value dramatic means of working through issues arising from abusive experiences (Van-der-Kolk: 1995/1997, Holmes: 2001 Batmanghelidjh: 2006, Cattanach: 2008, Lahad: 2009 – See Appendix 1).

Psychotherapy and the Arts Therapies have long been used to treat difficulties with attachment (Fonagy: 1997, Holmes: 2001, Meekums: 2005, Meldrum: 2007), as have narrative, drama and story, enabling exploration of personal historical events and the sense of self in a safe manner, been used in trauma work (Rothschild: 2000, Cyrulnik: 2005, Bannister: 2003, Cattanach: 2008, Lahad: 2009, Jennings: 2003/2010). I have presented evidence that Psychotherapeutic practitioners in the field of trauma (Batmanghelidjh: 2006, Sinason: 2002, Fonagy: 1997, Bloom: 1996, Holmes: 2001) acknowledge that abuse and insecure attachment are linked and that this combination may predispose sufferers to experience paranoia and undergo hallucinations. They have been shown also to advocate the use of therapies which come under the umbrella of dramatherapy work (Bloom: 1996, Goff and Goff: 1989, Wier-Perry: 2006, Holmes: 2001, Batmanghelidjh: 2006 – see Appendix 1).

I have reviewed the work of Dramatherapists working in the area of trauma and these have been shown to identify a link with attachment difficulties (Bannister: 2003, Gersie: 1997, Read-Johnson: 2009) and Attachment Theory has been revealed as providing a theoretical basis both for trauma work and also in understanding some of the complex difficulties arising from traumatic experiences in childhood and disordered attachment in the Psychotic client group (Meldrum: 2007, Jennings: 2010).
Finally, through descriptions of the work of other Dramatherapists in the field (Grainger: 1995, Snow: 1996, Casson: 2004, Emunah: 1994, Dunne: 2003) and also looking at my

own case histories, I feel I have demonstrated the validity of applying the techniques of attachment-based trauma work to work with Psychosis and how addressing attachment issues, traumatic experiences, the resultant thought disorder and voice-hearing via Dramatherapy processes could be used in Psychosis.

I feel this that both the hypotheses that Schizophrenia could be viewed principally as a severe form of Post-Traumatic Stress Disorder and that attachment-based Dramatherapy should be the treatment of choice for this client group have both been shown to be valid, although further research into the latter may still be needed.

Proposals for future Dramatherapy work with the Psychotic client group.

As I have revealed (see Appendix) a range of Dramatherapy interventions have been identified, via the literature, research and articles which I have reviewed during this study, as useful and appropriate for working with clients experiencing trauma-based Psychosis. Theorists on trauma and dissociation identify various common features which have been shown to be also present in those enduring Psychosis; all of which I feel could be most effectively addressed by Dramatherapeutic means.

For example distressing voices/ hallucinative phenomena/ flashback and fantastical experiences could be addressed via theatrical means, thus bringing difficult personal events into consciousness in order to make sense of and achieve mastery over their past history and to also address current difficulties. A skilled Dramatherapist could avoid re-traumatising the client during this difficult work through use of the aesthetic distance and containing boundaries of metaphor or projective techniques, as suggested by Jeremy Holmes and Phil Thomas.

Linda Winn's work using guided imagery, sculpting and empty chair would, to my mind, also clearly be helpful in working through Psychotic experiences, as would monomyth work as advocated by Boris Cyrulnik, John Weir Perry and David Read Johnson or ritual enactment as recommended by Sandra Bloom and Mooli Lahad. Focus on spiritual matters is certainly advocated in recovery literature and clients report that spirituality gives meaning to their struggle.

Many of the difficulties clients encounter as a result of childhood trauma and its aftermath of disturbed attachment, such as dealing with emotional self-regulation and impulse control, reduction in empathic abilities, fears of closeness/ impaired ability to trust and diminished sense of self would be addressed by a supportive Dramatherapy group experience, incorporating neuro-developmental bodywork and stress reducing activities to promote emotional equilibrium and trust in the self initially, as recommended by Babette Rothschild, Bessel Van-der-Kolk, Ann Bannister, Sue Jennings, Jenny Pearson and John Casson.

Later on, once a sense of safety has been established, the group could work on repair of damaged attachment re-enabling capacities for trust, hope, empathy and creative thought, by means of narrative work as sponsored by Anne Cattanach, Alida Gersie, James Roose Evans and Pamela Dunne and then socio/ psychodramatic enactment, such as promoted by Renee Emunah and Augusto Boal.

Impairment of interpretative and reflective abilities, concrete/catastrophic thinking due to defensively reduced mental flexibility and lack of resilience to stress are also much referenced in trauma literature. Grainger recommends supportive Dramatherapy groupwork as a productive means of dealing with these also, through shared participation in the essential social processes of human experience, enabling review and modification of disordered ideation through the social mirror, improved self-esteem via witnessing for eachother, mutual support and in time development of role expansion and re-integration of split off parts of the self.

This is exciting new work and I look forward to further engagement in my own research as to its effectiveness within my new profession in future months and years.

Rosemary Kate Hughes

2009.

References:

Axline, V. M. (1971). Play Therapy. Ballantine.

Bannister, A. (1997). The Healing Drama: Psychodrama and Dramatherapy with Abused Children. Free Association Books.

Batmanghelidjh, C. (2006). Shattered Lives: Children who live with courage. Jessica Kingsley.

Bentall, R. (2004). Madness Explained: Psychosis and Human Nature. Penguin Books.

Berry, K. and Barrowclough C. (2008). Understanding Symptoms and Interpersonal Relationships in Psychosis. Journal of Mental Health.

Bloom, S. (1996). Bridging the Black Hole of Trauma. London Institute for Self Analysis.

Boal, A. (1995). The Rainbow of Desire: The Boal Method of Theatre and Therapy. Routledge.

Bowlby, J. and Ainsworth, M. (1965). Childcare and the Growth of Love. Penguin Books.

Casson, J. (2004). Drama, Psychotherapy and Psychosis: Dramatherapy and Psychodrama with People Who Hear Voices. Routledge.

Cattanach, A. (2008). Play Therapy with Abused Children. Jessica Kingsley.

Clarke, A. and Browne, W. (2006). Writing on Walls. Stop Gap Theatre Co.

Creswell, J. (1994). Research Design: Qualitative, Quantitative and Mixed Method Approaches. Sage.

Cyrulnik, B. (2005). Talking of Love on the Edge of a Precipice: How to overcome trauma and remake your life history. Penguin.

Dent-Brown, K. and Ruddy, R. (2006). Dramatherapy for Schizophrenia or Schizophrenia-like Illnesses. Cochrane Database.

Diagnostic and Statistical Manual of Mental Disorders, Fourth Edition. (2000). American Psychiatric Assoc.

Dunne, P. B. and Rand, H. (2003). Narradrama: Integrating Dramatherapy, Narrative and the Creative Arts. Concordia.

Emunah, R. (1994). Acting for Real: Dramatherapy Process, Technique and Performance. Routledge.

Fonagy, P. (1997) Multiple Voices and Meta-Cognition. Journal of Psychotherapy Integration.

French, P. and Stevens, H. (2009) The Inter-relationships Between, Trauma, Dissociation, Cognition and Psychosis. British Journal of Clinical Psychology.

Gersie, A. (1997) Dramatic Approaches to Brief Therapy. Jessica Kingsley.

Gersie, A. (1997) Reflections on Therapeutic Storymaking: Like a Piece of Uncast Wood. Jessica Kingsley

Gerhardt, S. (2004) Why Love Matters. Routledge.

Grainger, R. (1995) Drama and Healing: The Roots of Drama Therapy. Jessica Kingsley.

Grof, S. and Grof, C. (1989) Spiritual Emergency: When personal transformation becomes a crisis. Warner.

Guba, E. G. and Lincoln, Y. S. (1992) Competing Paradigms in Qualitative Research. Handbook of qualitative research. Sage.

Gumley, A. (2009) Attachment-based Understanding: Engagement and Recovery after Psychosis. Psychological Medicine.

Holmes, J. (2001) The Search for the Secure Base: Attachment Theory and Psychotherapy. Routledge.

Jennings, S. E. (1978) Remedial Drama. Theatre Arts Books

Jennings, S. E. et al (1994) The Handbook of Dramatherapy. Routledge.

Jennings, S. E. (2010) Dramatic-Play and Healthy Attachments. Jessica Kingsley

Jung, C. G. (1978) Man and his Symbols. Picador.

Jung, C. G and Chodorow, J. A (1997) Active Imagination. Princeton.

Larkin, M. and Bloom, S. L. (1996). Dissociation and the Fragmentary Nature of Traumatic Memory. Journal of Psychotherapy.

Lahad, M. (1993). A Safe Place, Community Stress Prevention Center (CSPC) Publ.

Langley, D. M. and Langley, G. (1983). Dramatherapy and Psychiatry. Croom Helm.

Linford Rees, W. L. (1970). A Short Textbook of Psychiatry. Unibooks.

May, R. (2006). Working Outside the Diagnostic Frame. The Psychologist.

Meekums, B. (2005). Creative writing as a tool for assessment: Implications for embodied working, The Arts in Psychotherapy.
Meldrum, B. (2007). The Drama of Attachment. The Prompt. British Association of Dramatherapists.

Morgan A. et al. (2008). Being Human: Reflections on mental distress in society. PCCS Books.

Morgan, C. and Fisher, H. (2007). Environmental Factors in Schizophrenia. Schizophrenia Bulletin.

Moreno, J. L. (1993). Who Shall Survive. American Society of Group Psychotherapy.

Morrison, A. P., Frame, L. and Larkin, W. (2003). Relationships between Trauma and Psychosis. British Journal of Clinical Psychology.

N.I.C.E. (2009). Core interventions in the treatment and management of schizophrenia in adults in primary and secondary care. Dept. of Health.

Patten, M. (1980). Qualitative Evolution and Research Methods. Sage Publ.

Pearson, J. (1996). Discovering the Self Through Drama and Movement: The Sesame Approach. Jessica Kingsley.

Perry, B. D. (2001). Violence and Childhood: How Persisting Fear Can Alter the Developing Child's Brain. Textbook of child and adolescent forensic psychiatry. American Psychiatric Press.

Phillips, A. (2007) Going Sane. Harper Publ.

Read-Johnson, D. (1981). Dramatherapy and the Schizophrenic Condition. Drama in Therapy. Drama Publications.

Read-Johnson, D. and Emunah, R. (2009). Current Approaches in Drama Therapy. Charles C. Thomas Publications Ltd.

Read, J. and Gumley, A. I. (2008). Can Attachment Theory Help Explain the Relationship Between Childhood Adversity and Psychosis? New Directions.

Read, J., Mosher, L. R. and Bentall, R. P. (2004). Models of Madness. Routledge.

Read J, Perry BD, Moskowitz A, Connolly J. (2001). The contribution of early traumatic events to schizophrenia. Journal of Psychiatry.

Rebillot, P. and Kay, M.(1993). The Call to Adventure: Bringing the Hero's Journey to Daily Life. Harper Collins.

Roberts, G. and Wolfson, P. (2004). The Rediscovery of Recovery. Advances in Psychiatric Treatment.

Rogers, C. (1978). On Personal Power. Psychology Today.

Romme. M. and Escher, S. (1989). Subjective Experiences of Schizophrenia and Related Disorders: Hearing Voices - Implications for Understanding and Treatment. Intervoice.

Rothschild, B. (2000). The Body Remembers: Unifying Methods and Models in the Treatment of Trauma and PTSD. Norton & Co.

Schattner, G and Courtney, R. (1981). Drama in Therapy: Children. Drama Publications.

Schattner, G and Courtney, R. (1981) Drama in Therapy: Adults. Drama Publ.

Schoop, T. (1974) Won't you join the dance? Mayfield.

Sinason, V. (2002) Attachment, Trauma and Multiplicity: Working with Dissociative Identity Disorder. Routledge.

Stafford-Clark, D. and Black, D. (1984) Psychiatry for students. Unwin Hyman

Thomas, P. and Leudar, I. (2000) Voices of Reason, Voices of Insanity: Studies of Verbal Hallucinations. Routledge.

Van-der-Kolk, B. (1995) Dissociation and the fragmentary nature of traumatic memories: An overview and exploratory study. Journal of Traumatic Stress

Van-der-Kolk, B. (1997) Developmental Trauma Disorder. Journal of Traumatic Stress.

Weir Perry, J. (2006) Schizophrenia and The Hero's Journey. Berkeley University Press.

Willis, J. W. (2007) Foundations of Qualitative Research. Sage Publ. Inc.

Winn, L. (1994) Post Traumatic Stress and Dramatherapy: Treatment and risk reduction. Jessica Kingsley.

Yotis, L. (2007) Review of Dramatherapy Research in Schizophrenia: Methodologies and Outcomes. University Mental Health Research Institute. Routledge.

Appendix:

Breakdown of findings on areas of Dramatherapy work used with trauma/ Psychosis.

Winn: narrative, metaphor, mythwork, story, guided imagery, ritual, sculpting, projectve art, enactment and empty chair work.

Van der Kolk- theatre work, yoga, breathing exercises, roleplay, balance, music and movement .

Rothschild – bodywork, stress management/relaxation, balance, communicative exercises, myth, gestalt and psychodynamic work.

Sherborne – developmental movement, trust exercises and interactive games.

Schoop - body-ego technique, dance, movement, mime and roleplay.

Batmanghelidjh - creative and arts psychotherapies.

Cyrulnik – sociodrama, story, cultural and mono-myth groupwork, dreams and fantasy enactment, co-operative games.

Holmes - creative exploration of childhood issues, particularly using art, creative writing, poetry, claywork, story.

Bannister – non-verbal, psychodrama, embodiment, story development using metaphor, myth and symbolism.

Gersie - stories and the therapeutic group as a 'good enough' therapeutic base.

Cattanach - projective art, claywork, sandtray/ the Lowenfeld World Technique, images and narrative storywork

Lahad - art, drama, gestalt, narrative, shamanistic ritual nature and body–mind.

Read-Johnson – mythwork, embodiment, encounter and playspace

Jennings – EPR, embodiment, sensory and dramatic play, use of masks, sociodrama, quest/ mythical story enactment, sandtray work, clay, plaster and wood carving, poetry, ritualistic storytelling and dance.

Bloom - rite of passage work, music, dance , story,enactment of myths, using the metaphor of journey.

Armstrong - vocal work, rhythm, sound, chant, song, improvisation.

Pearson – bodywork, touch, push and pull, bridgework, story enactment.

Bergman – balance, breathing, Bobath bilateral exercises, mirroring, musical expression and creative movement.

Win-Browne - creative writing, storytelling and drama using fictitious characters, sociodrama, videowork, music, poetry, visual artwork, phototherapy, puppetry, work with voices and spectograms.

Romme and Escher - creative writing, meditation and ritual.

Emunah - dramatic play, improvisational and interactive exercises, theatre games, sociodramatic scenework, improvising dramatic scenes

Boal – sociodrama, psychodrama, roleplay, physicalisation of voices.

Dunne – narradrama, puppetry, poetry, art, spectogram work.

Meldrum - narrative, dramatic and projective work

Casson - EPR, relaxation, Tai Chi, voicework, imagined landscapes, spectogram, world techniques, storytelling, journey maps, ritual, myth, work with texts, poetry, puppetry, videowork, visualisation and enactment, photo therapy, mask and costume, brief psychodrama, work with voices and dreams.

Grainger - communication/ interactive games, role-play, scene enactment and shared physical exercises.

Snow – ritual, myth, fantasy enactment and Rebillot's heros journey.

Roose-Evans - creative art and journalling alongside rite of passage work.

www.ingramcontent.com/pod-product-compliance
Lightning Source LLC
Chambersburg PA
CBHW072302170526
45158CB00003BA/1153